Smart
Sugars

Sugars that Speak.
Why we should listen!

JC Spencer

WORLDWIDE PUBLISHING GROUP
Your Multi-Platform Publishing Partner
(713) 766-4272

An introduction to Glycoscience. Easy to read for the student while packed with new information for the seasoned medical professional, research scientist, and learned professor.

EBook 978-1-312-10589-8
Softcover   978-1-365-80901-9
Hardcover 978-1-312-10588-1

Table of Contents

Smart Sugars
Have inside information which can change your life.

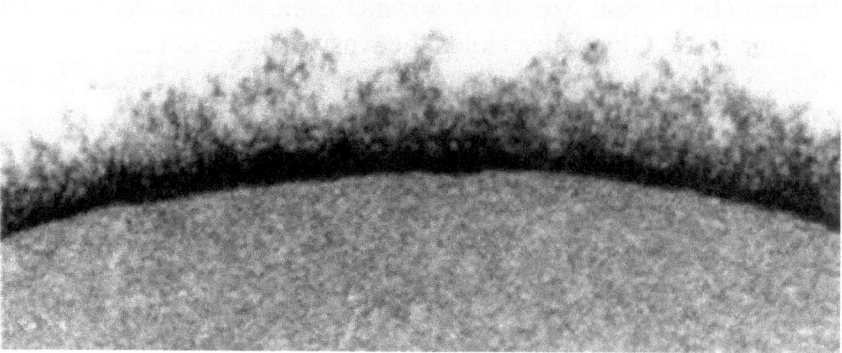

GLYCOSCIENCE
aka Glycomics and Glycobiology
S C I E N C E   O F   S U G A R S

About the picture
Glycoprotein receptor sites coat the surface of a red blood cell.
This glycan layer shows the physical structure of the
operating system (OS) of all cellular communication.
Source: National Research Council of the National Academes
Transforming Glycoscience - A Roadmap for the Future
supplied by Voet and Voet, Biochemistry, John Wiley and Sons, Inc. and used by permission.

Glycoscience is now proven to be the bull's eye, the Rosetta Stone, the Holy Grail of medicine and of all healthcare.

The author explains why in words easy to understand.

About the Author

Author JC Spencer condenses the knowledge of Glycoscience he has gained during the last two decades into three chapters for the layperson to better understand and for seasoned medical scientists, professors, and researchers to gain new information to better comprehend the depth and importance of this emerging discipline of science. Glycoscience, the subject of this book, is now proven to be the bull's eye, the Rosetta Stone, the Holy Grail, of medicine and of all healthcare. Everything points to accelerated and expanded glycoprotein research, and rightfully so.

He is CEO of The Endowment for Medical Research, Inc, a 501(c)(3) non-profit faith-based medical research and education public charity and think tank based in Houston, Texas, which conducts nutritional surveys throughout the United States, Canada, and some foreign countries.

For details on booking the author for lectures at universities and fund raising events contact him at jcs@endowmentmed.org

Preface

Most sugars and sweeteners are either harmful to human health or have no health benefit. Smart Sugars have health benefits and some Smart Sugars are downright ingenuous.

Scientists have discovered that Smart Sugars are vital to LIFE. They provide and may actually determine the quality of LIFE. The study of Glycoscience or Glycomics (smart sugars) is the emerging science that will change the way we live.

The Endowment for Medical Research, Inc., is a non-profit scientific research and educational public charity, and is committed to education of this science and to provide that education to healthcare professionals and the general public worldwide. This educa-tional effort, including Smart Sugars, was encouraged by the National Research Council of the National Academies. Major agencies of the U.S. Government recognize the major benefits of "Smart Sugars" and recently formed a committee to assess the importance and impact of Glycomics and Glycosciences.

In late 2012, the project, Transforming Glycoscience - A Roadmap for the Future, was approved and published by the Governing Board of the National Research Council, whose members were drawn from the councils of the National Academy of Sciences, the National Academy of Engineering, and the

Institute of Medicine. The project was supported by the National Institutes of Health, the National Science Foundation, the Food and Drug Administration, the Howard Hughes Medical Institute, and the U.S. Department of Energy. The group was charged to "articulate a unified vision for the field on Glycoscience and Glycomics" and to "develop a roadmap with concrete research goals to significantly advance the field."

On the website www.GlycoscienceNews.com sponsored by The Endowment for Medical Research, Inc., you can continue to learn more reasons why and how Glycoscience is the science that will **"change the way we live."** Reports on the work of the committee and ongoing Glycomics research will keep our readers informed with an easy-to-understand format. It has been the author's mission for nearly two decades to help educate and make it easy for the public to understand Glycoscience. Glycoscience may be thousands of times more complex than the genome project.

The committee for the project, <u>Transforming Glycoscience - A Roadmap for the Future</u>,

CONCLUDED that
integrating Glycoscience
into relevant disciplines in
high school, undergraduate,

and graduate education,
and developing curricula
and standardized testing
for science competency
will increase public as well
as professional awareness.

National Research Council of the National Academies <u>Transforming Glycoscience - A Roadmap For The Future</u> Summary page 12

The future of medicine and healthcare is Glycoscience.

Like thick fuzz on a peach, healthy cells are coated with these Smart Sugars.

Medical scientists
agree that the Gold
Standard for future
medical diagnostics,
especially for cancer,
will be to analyze the
quantity and quality
of these sugar
molecules on the
surface of cells.

Introduction

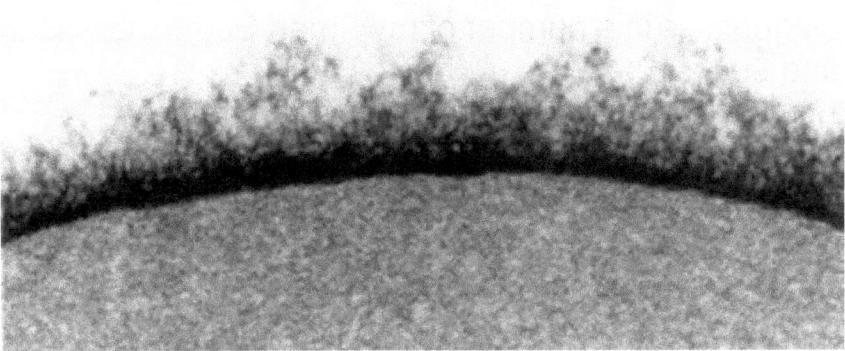

You are looking at glycoprotein receptor sites on the surface of a red blood cell. This glycan layer is more than just the operating system (OS) of all cellular communication for the body. The Smart Sugars, the building blocks of the glycan layer, give LIFE to the cell.

Smart Sugars are the foundation of all LIFE on Earth. *"The most important biochemical process on Earth is photosynthesis ... using energy in sunlight to combine carbon dioxide and water to make sugars."*[1]

Like thick fuzz on a peach, healthy cells are coated with these Smart Sugars. Unhealthy cells have less dense glycans or, as on cancer cells, the polysacchrides are diminished or missing altogether, leaving the cells bald, toxic, and infectious. Medical scientists agree that the Gold Standard for future medical

13

diagnostics, especially for cancer, will be to analyze the quantity and quality of these sugar molecules. I discuss the emerging technology of glycoprotein diagnostics in a number of my Smart Sugars Lessons and books.

The purpose of <u>Smart Sugars</u> is to provide a starting point for all people to learn about the imminent paradigm shift in medicine and health care. Super Sugars are evidence-based and the results are in the irrefutable facts. The emerging data from ongoing research are nothing short of amazing.

The future of medicine and healthcare is Glyco-science. Undeniably, scientists agree that the ex-plosive technological advances in Glycoscience will transform medicine and healthcare. Glycoscience has the potential to dramatically reduce health care costs.

Glycoscience is rapidly becoming a part of main-stream medicine and was declared by Massachusetts Institute of Technology (MIT) as

"ONE OF THE 10 EMERGING TECHNOLOGIES THAT WILL CHANGE THE WORLD."[2]

Working together, The National Institutes of Health (NIH), the Food and Drug Administration (FDA), the National Science Foundation (NSF), the National Re-search Council, the National Academy of Sciences, the Howard Hughes Medical Institute, and the U.S. Department of Energy (DOE), formed a committee to evaluate the importance, impact, and future of Glycoscience. The group was charged to "articulate a unified vision for the field on Glycoscience and Glycomics" and to "develop a roadmap with concrete research goals to significantly advance the field." We will report on the work of this committee and keep our readers informed with an easy to understand format.

It is exciting for these significant agencies to request Glycoscience be taught in high school science classes as well as undergraduate and graduate education. This will assure that future generations will know about the science.[3]

Our website, **www.GlycoscienceNEWS.com**, will continue to provide a developing vision of the Glycoscience of tomorrow and how the future of the human race will be enriched in ways only dreamed of before. On this site, we will report where Glycoscience can take us today and tomorrow.

Four FACTS about Smart Sugars

Most Smart Sugars are:
(1) unknown to the public,
(2) not sweet (some are sweet),
(3) extremely beneficial to good health, and
(4) the building blocks for the operating system (OS) of the human body.

These specific sugars, often called glyconutrients, are unique. Contrary to common thought, glyconutrients are functional beyond just supplying energy to the body. Your body uses these sugar building blocks to construct the actual operating system (OS) within (glycolipids) and on the surface of (glycoproteins) every cell.

The sugars and sweeteners consumed by Westerners, such as table sugar (sucrose) and synthetic sweeteners, lower the quality of health and contribute to obesity, diabetes, and other diseases plaguing humans. Bad sugars and sweeteners weaken the immune system and all the other vital functions of the body. The smartest of the really good sugars can actually help modulate your immune system and balance your hormones which is vital to your quality of life and longevity.

Prior to fifty or so years ago, our ancestors ate more unprocessed foods and natural sweeteners containing Smart Sugars. Today our foods are depleted of many nutrients and necessary Smart Sugars. It is more important today than ever before to supplement our diets with these beneficial, functional sugars along with natural vitamins and minerals.

The answer to the healthcare crisis is to maintain good health through prevention. A strong, well modulated immune system and well functioning endocrine system are vital to your health and longevity. Smart Sugars may help make that a reality. Insulin is a hormone that can upset the immune and endocrine systems.

[1] Transforming Glycoscience - A Roadmap For The Future

Introduction page 13.

[2] MIT Technology Review, 10 Breakthrough Technologies That Will Change the World (Glycomics) February 2003

[3] National Research Council of the National Academies Transforming Glycoscience - A Roadmap For The Future Summary page 12

**The answer
to the healthcare crisis
is to stay well.**

The question is how will Smart Sugars help us stay well.

Chapter One

What Are Smart Sugars?

The children of today are not expected to live as long as their parents lived nor be as healthy as their parents.

The children of today are not expected to live as long as their parents lived nor be as healthy as their parents.[1]

This is alarming!

Paradoxically, discoveries in Glycoscience suggest it is possible to live in good health longer than previous generations.

Premature aging and the diseases associated with aging are appearing in our children in unprecedented numbers. Paradoxically, new discoveries in Glycoscience suggest it is possible to live in good health longer than previous genera-tions.

Everyone wants to be healthy, but once people face serious health challenges, it is often too late to reverse the effects of years of aging and damage at the cellular level. Therefore, the earlier in life good health habits are instilled, the better long-term quality of life a person should have.

The news is just starting to get out about these super healthy sugars.

Anti-aging researchers, including George S. Roth, Ph.D., believe we can live a vibrant, healthy life to the age of 150 if we make the right choices, eat right, maintain a high pH, exercise properly and keep our minds alert. Dr. Roth, who is author of The Truth About Aging and former researcher for the National Institute on Aging, is optimistic, "*In addition to eliminating a lot of the diseases of aging, we'll maintain function, vitality, cognition and all the other things we value in terms of quality of life.*" In March 2001, *Science* magazine dedicated twelve articles (3/2001) to educating its science and medical readers about functional sugars.

In the 1980s, Bill McAnalley, Ph.D., a pharmacologist and toxicologist in Texas, discovered the first known sugar with medicinal benefits, the beta-mannan molecule. Earlier research conducted by Clarence "Lush" Lushbaugh, M.D., Ph.D., for the Atomic Energy Commission in 1952, pointed Dr. McAnalley toward the beta-mannan molecule.

The beta-mannan sugar molecule is a phytonutrient found in certain plants. It is the functional component in Aloe Vera but is destroyed shortly after the leaf is harvested. The fact that an enzyme begins breaking down the long chain mucilaginous polysaccharide beta-mannan sugar molecule resulted in many contradictory published papers. Some researchers were using fresh Aloe Vera that proved beneficial and some were using leaves that were a few hours or days old that had no beneficial effect.

The term, "glycobiology" was coined in 1988 at Oxford University by Professor Raymond Dwek.

In the 1980s, Bill McAnalley began his search to isolate the functional component in Aloe Vera after studying a paper written by Clarence C Lushbaugh, M.D., Ph.D. In 1949, Lush, as he liked to be called, worked as a patholo-gist at the Los Alamos Medical Center in New Mexico where he served as a staff member in the Biomedical Research Group of the Los Alamos Scientific Laboratory.

Dr. Lushbaugh was a path-finder in forensic pathol-ogy. He studied the rate of body cooling to use in estimating the time of death and published a report of his findings in a law enforcement publication that became known as the Lushbaugh method. Lush authored or co-authored more than 150 scientific papers and book chapters.

The Atomic Energy Commission was seeking a way to develop something that could be stored in fallout shelters to treat radiation burns in humans. Dr. Lushbaugh conducted research on experimental animals with his focus on biochemical changes in irradiated skin. His rabbits concealed a valuable secret until Dr. McAnalley solved the mystery years later after reading his papers.

In 1952, having become internationally known as an authority on radiation accident victims, he tested Aloe Vera on rabbits with six beta radiation burns. He placed a fresh fillet of Aloe Vera on two of the burns, week-old Aloe Vera on two of the burns and the other two burns were the control burns that received no treatment. To his amazement, the two beta radiation burns treated with fresh Aloe Vera fillet healed completely to new rabbit pink skin. The two burns treated with the week-old Aloe Vera did no better than the two control burns that received no treatment.

The Atomic Energy Commission was fearful of World War III. Although Dr. Lushbaugh found the solution for treating radiation burns, he did not know how to preserve it. More than thirty-five years later, Dr. McAnalley, after reading Lush's report, was attempting to discover what was the functional component of aloe vera, thinking it was a protein.

Surprising connections brought about the great discovery. It happened when Bill "Nuked" Aloe Vera.

The Atomic Energy Commission
Clarence C Lushbaugh, M.D., Ph.D.
Bill McAnalley, Ph.D.

One day, almost on a whim, Dr. McAnalley placed some fresh Aloe Vera in a microwave and "nuked" it. The message the "nuked" Aloe delivered to Dr. McAnalley was that the functional component was, indeed, not a protein, but a carbohydrate, a sugar. He knew if the functional component were a protein, it would have been destroyed by the microwave. But, it was not destroyed. It was still functional. Because the beta-mannan molecule was a sugar, the medical world scoffed at the discovery of a "beneficial sugar pill". Medical scientists, steeped in pharmaceutical education, could not accept the fact that a sugar had any functionality other than for storing energy.

It has been a long time coming, but today, the mainstream medical community is beginning to grasp the profound validity of Glycoscience. More than half a century after Dr. Lushbaugh's paper, Glycomics is the new frontier of medical science. Dr. Lushbaugh died at age 84 on October 13, 2000, from Alzheimer's.

Nearly 650,000 references for "glycoprotein" are cited by The National Library of Medicine. Scientists are adding so many new research papers each day (thousands per month) that no one can adequately keep up with the progress of Glycomics.

How these sugars bond is critical to their function. All successful systems utilize a structure design and have a function. Chaos happens when the design and function do not fit the system.[2] The system of the cell is corrupt without Smart

Sugars. Humans **unknow-ingly** depend on these sugars for life for cell to cell communication through complex electrical and chemical signals. Smart Sugars read and transcribe DNA and make the determination of blood type.

Glycoscience holds the key for us to live a longer, better quality, vibrant life.

An exciting facet of glycobiology is found in the blood. While there are four basic blood types: A, B, AB, and O, there are thirty human blood group systems and 600 known antigens that characterize the proteins discovered on a human red blood cell. Each of these thirty blood types is a classifi-cation of blood based on the presence or absence of inherited antigenic substances on the surface of the red blood cells. These antigens may be proteins, sugars (glyco), glycoproteins, or glycolipids. How these sugars are arranged determines the blood

These Smart Sugars even determine your blood type.

type. One sugar that determines the blood type is so important that when another blood type is used in a transfusion, the patient dies.

A fact that is wonderfully amazing to me is that Smart Sugar glyconutrients are abundant in breast milk and the new born baby needs these for preventing infections. Research papers have documented that children who received glyconutrients from breast milk were healthier, have a superior immune system for years.[3] There is evidence that students make better test scores in college than those who did not receive these sugars when they were children.[4]

Oxford University advanced the science of sugars, and in 1988, Oxford Professor Raymond Dwek coined the term "glycobiology", which is widely used today. Scientists have identified about twenty seven Smart Sugars found in nature. Here is a brief look at five of these Smart Sugars. Some sugars may appear smarter than others but remember, we still don't know them very well. They have been speaking for a long time but, we have just started listening.

- **mannose:** Studies show that mannose has remarkable health benefits, especially involving the immune system, cognitive functions, and cancer. Humans are dependant upon it. Mannose is cited by the National Library of Medicine in more than 25,000 references linked to research.
                (Reference: NIH   public website,

www.PubMed.gov)

- **fucose:** (not to be confused with fructose) Fucose, with major health benefits, may prevent and treat cancer, and the National Library of Medicine cites more than 9,000 references with more than 1,200 links to studies in cancer research.

(Reference: NIH public website,

www.PubMed.gov)

- **trehalose:** The unique bond of two glucose molecules. Trehalose is able to protect the integrity of cells against a variety of environmental stresses such as desiccation, dehydration, heat, cold and oxidation. It is a non-reducing sugar with its non-reducing end to the left that is not easily hydrolyzed by acid. Trehalose has a functionality that aids in the proper folding of proteins. More than 5,400 references are cited by the National Library of Medicine links to research.
(Reference: NIH public website,

www.PubMed.gov)
- **galactose:** Studies show that galactose has health benefits, humans are dependant upon it especially babies to build their immune system and help structure the glycoprotein receptor sites. More than 31,000 references cited by the National Library of Medicine are linked to research.
(Reference: NIH public website,
www.PubMed.gov)

- **glucose:** The medical establishment recognizes the basic sugar glucose as very important to human life. However, it appears to

be the most harmful in large quantities, especially for
diabetics.   Glucose is used in solution to treat nearly everyone in hospitals.
(Reference: NIH public website,
www.PubMed.gov)

Drug companies are rushing to synthesize these sugars into new drugs.

Researchers are giving the class of Smart Sugars serious explor-ation. More physicians are beginning to incorporate these sugars into their practice by asking their patients to eat them. More individuals are ingesting these natural sugars and discovering health benefits. Drug companies are rushing to synthesize these sugars into new drugs. Many thousands of patents related to these super sugars have been issued, most since 1995.

We are presently aware of a small but growing number of very significant super sugars. Eight of these sugars were presented by Robert K Murray, MD, Ph.D., and published in multiple editions of Harper's Biochemistry.

Without these Smart Sugars the human cell would have no life.

It has become self-evident and scientists have concluded that super sugars have efficacy individually and synergistically to build cells, coat cells, strengthen cells and cell membrane. Smart Sugars become proficient with other cells to form the whole neurological mental and motor systems that operate the human body.

These super sugars provide quality of life to the human body.

In short, these super sugars provide quality of life. Without them, you would have no life.[5] Some scientists have debated whether your body can produce all of these sugars without ingesting them if you initially only have glucose in your body. It is theoretically correct that

Glycomics research is ongoing in universities around the world.

your body could manufacture the other sugars from just one. It is a fact, however, that while your body has the potential ability to manufacture them, your body does not manufac-ture enough because of the time and energy required and the enzymatic gymnastics necessary.

When we do not eat what our bodies requires, the necessary nutrients are robbed from our body parts to make up for the need. This lack may also be the reason for cravings of sugar or something sweet to take up the slack.

Some of these sugars are very expensive in pure form. One is upward to $4,000/ kilogram and another nearly $25,000/kilogram. Glycomics research is ongoing in universities around the world. The

Endowment for Medical Research has supplied sugars to universities and research labs in at least six countries and conducted Pilot Surveys in three countries.

Here is a partial list of significant sugars. While I have listed twenty seven sugars here, there may be many more unique sugar structures yet to be discovered. An ever-so-slight change in the sugar structure provides a unique functionality that may distinguish it as a new sugar with a unique function. A more exhaustive study of these sugars can be found in my other books and available online on our websites.

An ever-so-slight change in the sugar structure provides a unique functionality that may distinguish it as a new sugar with a unique function.

The list of Smart Sugars below is not exhaustive and is not in alphabetical order.

- fucose (not to be confused with fructose)
- L-fucose   • GDP-fucose • D-fucose
- mannose • D-mannose • glucose
- arabinose • D-arabinose • trehalose
- GDP-mannose • ribose   • rhamnose
- GDP-alpha-D-mannose • L-rhamnose
- glucosamine • galactosamine • xylose
- n-acetylgalactosamine   • galactose

- melibiose
- lactose
- sialyllactose
- n-acetylglucosamine
- D-ribose
- n-acetyl-d-galactosamine
- n-acetylneuraminic acid

Families are finding that trehalose sugar is probably easiest to use of all of the Smart Sugars because unlike most functional sugars which are not sweet, this sugar has a delightful sweet taste. Families simply exchange what is in their sugar bowl for trehalose. It looks like regular table sugar and pours better than regular table sugar. While not quite as sweet, it has a more pleasant, clean taste with no aftertaste.

The news is just starting to get out about these super healthy sugars and you are some of the first to know. Discovering and using Smart Sugars can help improve long-term wellness for us and that of our children and grandchildren. Glycoscience holds the key for us to live a longer, better quality, vibrant life.

Because all of us are most likely deficient in most of the Smart Sugars, my wife and I personally consume a variety of glyconutrients daily and believe that we are the healthier for doing so. My wife (65 in 2013) (she gave me permission) and I (74 in 2013) are on no medication and never expect to be. I have more energy and especially seem to be more mentally alert from eating Smart Sugars every day.

Our webmaster, Jim Wing was 439 pounds in 2006. Jim had a near death experience in 2012 because of massive glucose imbalance (tested 770 sugar load).

He is now 63 years old and the doctors are amazed that his vital signs, organs, and arteries are excellent. He is now fully committed to taking solid sound scientific action towards fat loss while gaining better health in the process. We believe good nutrition and Smart Sugars saved Jim's life. He learned from this dramatic event that it's not good to dump junk food into the feeding trough.

Jim is on his journey to renewed health. He is already able to back off much of his insulin by maintaining a consistently low glucose level, and is working on lowering his triglycerides. He learned that high triglycerides cause chaos with the transporting and processing of fat and glucose in and out of the liver.

How these sugars bond is critical to their function.

His doctors are astounded that he has recovered and improved so quickly with consistency of his tight low sugar count. They said, *"This does not happen!"* They will continue to closely monitor his progress.

Functional sugars can be purchased from nutritional companies, online stores, or product distributors. But, buyers need to get the facts on each product about content and quality control.

A caution especially about aloe products: Remember the story of Dr. Lushbaugh? The functional mannose molecule is destroyed by enzymes shortly after the leaf is har-vested. Years ago, the FDA almost shut down the aloe industry because the functional mannose was missing in most aloe products they tested from health food store shelves. Purchase your Smart Sugars from a reputable company that has stabilized the functional sugar molecule and preferably from one with a money back guarantee.

References:

[1] (pg 18) A 2005 study published in the *New England Journal of Medicine* confirms that, given the current health trends, today's young people are not expected to outlive their parents for the first time in U.S. history.

[2] (pg 24) Chaos happens when... is explained in Chapters 7 and 11 of Expand Your MIND - Improve Your BRAIN.

[3] (pg 25) *Science & Medicine* Vol 4, Number 6, November/December 1997 *Breastfeeding Stimulates the Infant Immune System* by Lars Å Hanson; Telethon Institute for Child Health Research in Perth, Australia published in January 2011 issue of the journal *Pediatrics*.

[4] (pg 25) American University and the University of Colorado Denver study Published in *Journal of Human Capital* Vol. 3, No.1, Spring 2009.

[5] Transforming Glycoscience - A Roadmap for the Future, by National Research Council of the National Academies. Introduction pg 15 - Important Facts About Glycans, Health 1. "*Elimination of any single major class of glycans from an organism results in death.*"

Look Back
on Chapter One

What the author calls, "Smart Sugars" are the sugars found in nature that are beneficial to human health.

Some of the Smart Sugars

are the building blocks of the operating system (OS) that process all data used in cell to cell communication. These sugars read and transcribe the DNA, determine your blood type, and play vital roles in programming your brain.

Look Forward on Chapter Two

Smart Sugars hold the key to the Holy Grail of medicine and all healthcare for future diagnostics, prevention, and treatment.

Most FDA-approved cancer biomarkers are already glycoproteins.

Chapter Two

Glycomics Holds the Answer to Cancer

for Prevention, Diagnosis, and Treatment

Glycoscience is now proven to be the bull's eye, the Rosetta Stone, the Holy Grail, of medicine and of all healthcare. Everything points to accelerated and expanded glyco-protein research and rightfully so.

The report states, "Most FDA-approved cancer bio-markers are glycoproteins."

In 2008, I reported on developments for measuring glycoproteins which would become the Gold Standard for medical diagnostics. *Science Daily* published a story of detecting early forms of cancer by analyzing the structure of specific sugar molecules. Another report revealed "A Serum Glycomics Approach to Breast Cancer Biomarkers".

A NIH Glycomics Cancer study through NCI (National Cancer Institute) in cooperation with the Alliance of Glycobiologists and various universities was initiated in 2007. Funding was $32 million per year for a number of years. Today we look at the conclusion of that research.

The report states, "Most FDA-approved cancer bio-markers are glycoproteins", but little is known about how their glycan structures are altered in cancer where diagnostic performance could be greatly improved. This experimental data is available at MIT.

The purpose for this study was to attract qualified scientists to exploit fundamental aspects of cancer biology; and, to establish a core of integrated glyco-biologists to facilitate the development of cancer glycobiology for presentation and diagnostic applications.

One point in the report said that incentives are needed to attract the leading glycobiologist experts to do cancer research with defined clinical goals. The problem, according to the report, is that traditional funding mechanisms are not suited for an emerging field.

"The Mission of the NIH/NCI study was to elucidate the structure and function of glycans that contribute to oncogenesis. And, to exploit aberrant glycosylation in cancer for the development of transla-tional applications for cancer prevention, detection, and diagnosis."

Translation: **"Why does poor glyco-sylation contribute to cancer and what can we do about it?"**

In 2007, Texas voters approved a $3 billion cancer research program. This has further equipped Houston as the hot bed of cancer research in the United States. The Houston Chronicle's front page headline on June 2, 2011 was "City Attracts Big Guns in Fight Against Cancer".

Universities around the world have generated conclusive data that animals and humans that consume Smart Sugars have phenomenal health benefits with both the immune system and brain function.

This influx of talent is a game changer for fighting cancer. The best and brightest are coming to Texas and Houston. These established researchers are adding enormous talent to Baylor College of Medicine, Methodist Research Institute, Rice University, University of Texas MD Anderson Cancer Center, University of Texas - Austin, San Antonio and Southwestern Medical Center.

Houston is the Medical Capitol of the world and MD Anderson is the leading cancer treatment center. A few years ago an oncologist at MD Anderson explained to me that a doctor friend in Boston told him that Glycomics was the future of medicine. MD Anderson's President Dr. Ronald DePinho calls the voter approved cancer program one of the factors that convinced him to leave Boston and come to Houston.

According to the NIH/NCI study, glycoproteins are good for diagnosing, monitoring, proving, reproving, testing, and researching to develop billions of dollars in drug sales. Perhaps in another ten to twenty years, we will have a cancer cure.

Measuring the quality and quantity of glyco-proteins on the cell surface is an excellent diagnostic approach because the lower the glycoprotein count, the greater the cancer risk. Healthy cells are sugar coated with glycoproteins while cancer cells are bald or balding. The misfolding of proteins is the cause for degrada-tion of the cell, resulting in poor quality and quantity of glycoproteins. Learning how to help the cells properly fold proteins is the future of medicine and health care.

Yes, glycoprotein technology IS the bull's eye but perhaps the target is placed over the wrong objective. The traditional medical target is still over symptoms using drug treatments, surgeries, or radiation. In the process, many are helped but, we can do so much more if we simply move the target from symptom to cause.

Over the last couple of decades, universities around the world have generated conclusive data that animals and humans that consume Smart Sugars have phenomnal health benefits with the immune system, brain function, and, directly or indirectly, virtually all diseases. Can we slash a trillion dollars out of healthcare cost by just eating these sugars? I believe we can.

> **Can we slash a trillion dollars out of healthcare cost by just eating these sugars? Perhaps we can.**

For a fraction of the billions invested in cancer research, we may be able to verify through pilot surveys and clinical studies what can be accomplished in humans with Smart Sugars and balanced body pH. One sugar, mannose, eradicated cancer in the American poultry industry. That story is documented in my Glycomics text-book, **Expand Your MIND - Improve Your BRAIN**. The National Cancer Institute verifies what I have been teaching for nearly two decades: The cause for cancer may be low quality and quantity of glycoproteins on the surface of cells. I applaud the NCI for their ability to determine this fact and to use Glycoscience as the ultimate focus for future diagnostics. The same Glycoscience diagnostics used to count these receptor sites should encourage us to grow more glycoprotein receptor sites through proper care and feeding of the cell.

One sugar, mannose, eradicated cancer in the American poultry industry!

What can all of the sugars together do?

> **Let your medicine be your food and your food be your medicine.**
>
> Hippocrates (460 - 377 BC)

The National Cancer Institute verifies what I have been teaching for nearly two decades: The cause for cancer may be low quality and quantity of glycoproteins on the surface of cells.

I applaud the National Cancer Institute for their ability to determine this fact and to use Glycoscience as the ultimate focus for future diagnostics.

Look Back on Chapter Two

Measuring and monitoring the condition of the glycoproteins on the surface of human cells will enable doctors to determine the present and future health challenges.

Glycoscience research programs are ongoing in universities around the world. Understanding the Smart Sugars will be the paradigm shift in medicine and all of healthcare.

Look Forward on Chapter Three

The author's wife says this next chapter made her brain hurt. This is for the scientist or science junky who dares to venture where no man has gone before... Quantum Glycoscience.

The relativity factor of future science is the understanding of entanglement.

Chapter Three

Glycoscience Meets Quantum Physics

This chapter is dedicated to the scientists at heart. My wife is not one of them. She thinks you will probably want to skip over this chapter that made her brain hurt.

Quantum science was a hard pill for Albert Einstein to swallow because of its mind bending, paradoxical explanation of phe-nomena. Today, a little bit of sugar will, that's right, help make the medicine go down.

Quantum biology is an emerging science discipline. You have probably never heard of Quantum Glyco-biology or Quantum Glycomics (QG). This chapter is your introduction and you will see how QG will help us solve the mysteries of why some sugars are so unbelievably beneficial for improving health and may in the future even be able to correct otherwise impossible devastating illnesses.

Scientists are grasping at quantum's bizarre proper-ties to solve mysteries of the evident influence of unseen forces. We will learn how to harness quantum influences; but, first we need to understand how the same wave-particle properties can produce drastically different outcomes.

I could hear you whisper, "*How can thought be involved in quantum glycobiology*?"

In QG, it will be necessary to understand how the folding of different proteins and sugars are entangled with unknown forces including ions, magnetism, photons, radiation, and thought. All this is manipulated further with variant thermal condi-tions, light of various spectrum, rate of radioactive decay, direction of rotation, speed of spin, angles, gravity, and the electrical discharge transfer of energy. Perhaps anything is possible with quantum mechanics.

The real relativity factor of future science is the understanding of entanglement.

I could hear you whisper, "*How can thought be involved in quantum glyco-biology*?" Think! In clinical studies, the placebo effect works on about a third of the patients who are taking a sugar pill, a "bad sugar" pill yet. Animals don't exhibit the placebo effect.

It was the variant factors of entanglement that baffled Einstein and caused him to call quantum physics, **The reason the entanglement factor is so important is the fact that it often works at the tipping point of efficacy.**
"*spooky action at a distance*". The real relativity factor of future science is the understanding of

entanglement. The reason the entanglement factor is so important is the fact that it often works at the tipping point level of efficacy. The tipping point factor can be explained with a perfectly balanced scale holding in each bucket half the water of all the oceans. The tipping point for the scale to tilt either direction is to add one drop of water to your choice of either side. This is how relative a drop of H^2O is when it puts its weight behind a purpose.

In quantum physics, the possibilities are endless, not just one tilting to the left or right like the drop of water on the scale. The endless possibilities of mysterious influences can alter the state and performance of a molecule or a system. The effect, the behavior, the consequences, are as pervasive as they are profound.

It is the proper folding of proteins that gives order to the human body and the misfolding of these same proteins that causes neuro-degenerative diseases.

Quantum influences are relative to all systems regardless of size. Quantum physicists have been concerned primarily with microscopic anomalies. There may be no boundaries of quantum prediction because of the incoherence of unknown factors. It is this entanglement that binds all the particles together to produce unknown quantities for strange conclusions. Here collective properties become impossible to untangle.

Each thought triggers a constellation of synapses in your brain to take flight like a flock of birds.

Like the tilted scale, when propensity is altered, that newly directed influence may gather momentum and influence com-pounding change. I postulate that the peer pressure of particle momentum influences atoms, ions, photons, magnetism, and thought. We have in previous lessons discussed how a thought triggers a constellation of synapses in your brain to take flight like a flock of birds. Scientists have learned that elementary particles also react like waves of activity and develop a propensity to operate in unison. This is the law of attraction manifest.

In other writings, I will discuss how it is the very entanglement that produces the outcome. It is mangled entanglement that produces chaos. It is the proper folding of proteins that gives order to the human body and the misfolding of these same proteins that causes neurodegenerative diseases.

Quantum glycobiology resides in the entanglement of proteins and sugars forming glycolipids and glyco-proteins that are the Operating System (OS) of the human body.

Vlatko Vedral of Oxford admits, "*Implications of macroscopic objects such as us being in quantum limbo is mind blowing enough that we physicists are still in an entangled state of confusion and wonder-ment* ".

It appears no one understands quantum physics yet, that has not kept brilliant minds from enjoying the possibilities nor kept them from babbling utter nonsense about things that are not relative. Quantum physics extends new opposing challenges to the theory of relativity. And, QG opens the door for under-standing the benefits of sugars like never before. Meanwhile, the efficacy of Smart Sugars is self-evident.

Quantum Glycoscience opens the door to understand sugars like never before.

Glossary

(Not in alphabetical order - categorically arranged)

Glyco: Greek for sugar.

Glycoscience: Science of sugars.

Glycomics: Science of sugars.

Glycobiology: Biology of saccharides, sugar
                    chains, and glycans.

Glycosylation: Process in which a sugar is
                    attached to another molecule, as
                    attached to a protein for them to
                    become glycoproteins.

Glycan: Sugar structures assembled one
                    sugar at a time as building blocks
                    to be linked and bonded usually to
                    a protein.

Glycoconjugates: Conjugated glycans (yoked
                    together)

Carbohydrate: Organic compound that includes
                    sugars, starches, and celluloses,
                    produced by photosynthetic plants
                    and contain only carbon,
                    hydrogen, and oxygen.

Sugar: A carbohydrate of carbon, hydrogen and oxygen. It is estimated that there are 200 various types of sugars found in nature.

Saccharide: Simple sugar molecule.

Monosaccharide: A carbohydrate that does not hydrolyze (maintains molecular structure), as glucose, fructose, or ribose.

Disaccharide: Two monosaccharides bonded together as in maltose, lactose, sucrose (invert sugar with alkaline stability), and trehalose. Each has a unique function determined by the bond.

Oligosaccharide: A carbohydrate made up of a few linked monosaccharides.

Polysaccharide: A complex carbohydrate composed of sugar molecules linked into a branched or chain structure.

Homopolysaccharides: Polysaccharides formed from only one type of monosaccharide subdivided into straight-chain and/or branched-chains.

Heteropolysaccharides: Polysaccharides formed from two or more different types of monosaccharides in straight-chain and/or branched-chains.

Glyconutrient: A food or supplement containing bioactive and functional sugars.

Phytonutrient: Phyto is Greek for plant Bioactive plant nutrient Functional sugars are phytonutrients but all phytonutrients are not sugars.

Phytochemical: Functional sugars are phytochemicals but all phytochemicals are not sugars.

Additional reading and training materials

Online sales support continuing Glycomics education and research.

Bookstore: www.endowmentmed.org

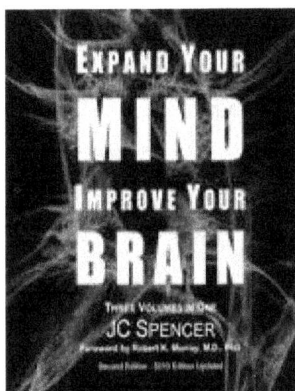

Expand Your MIND - Improve Your BRAIN is an easy to read entertaining science textbook. Over 700 M.D.s, Ph.D.s, Scientists, Researchers and Educators are referenced in the field of Glycomics and Brain Function. More than 500 page Textbook. First Edition published in 2008 Second Edition - 2013 Edition - Updated
Available as an e-textbook, perfect bound 8 1/5 x 11, and hardbound editions

Three Volumes in One
    Softbound .... $ 77.77
    Hardbound $127.77

Vol. 1; Vol. 2; and Vol. 3 available individually as ebook only .... $ 27.77 each
Three Volumes in One e-textbook $ 47.77
Available in the Book Store at
www.endowmentmed.org

51

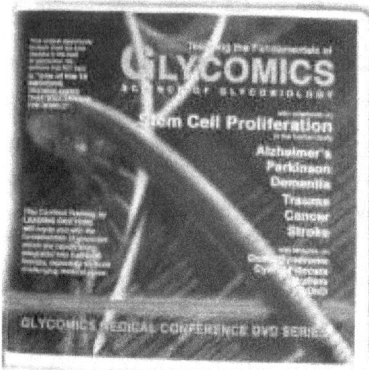

14 hours Professional Glycomics DVD Training Series from Glycomics Conference for Healthcare Professionals includes 500 page syllabus on CD of the color slides presented (SAVE $100 off regular price of $299) $199
(Testing available)

14 hours General Public Glycomics DVD Training Series from First Glycomics Conference for the General Public (does not include 500 page syllabus of all the color slides presented (SAVE $100 off regular price of $199) $99

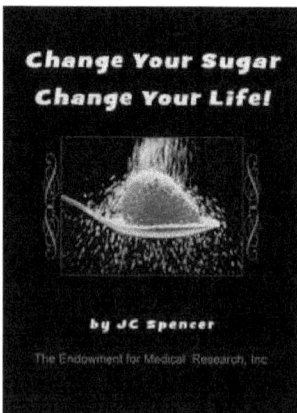

Change Your Sugar - Change Your Life!

This easy to read book is for the student and is packed with new information for the seasoned medical professional, research scientist, and learned professor.

The book explains the scores of diseases and health challenges caused by regular table sugar and what you can do about it.

Four Chapters include:
Chapter 1: CHANGE YOUR SUGAR, CHANGE YOUR LIFE
Chapter 2: THE SUGAR COMPARISON CHART
Chapter 3: GIVING SUGAR THE AROMA OF PURE HEALTH
Chapter 4: NEW STUDIES ON SUGARS AND CINNAMON

$27.77 softbound edition

Change Your Sugar, Change Your LIFE may be downloaded free of charge at

www.DiabeticHope.com

Stem Cell Survey

A CD Technical Syllabus

by H. Reg McDaniel, M.D. provided for use by Healthcare Professionals. Evidence that Glycomics can increase stem cell proliferation and stem cell function in humans. Learning and Behavior Problems in Children Responsive to Micronutrients Led to Benefits Reported in Infants and Youth and Maternal Alcohol Damage (FAS). Your contribution of $50 serves as a fundraiser and is shared between The Endowment for Medical Research and the Fisher Institute for Medical Research.

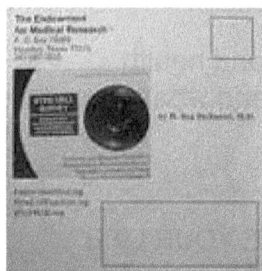

Additional reading
and training materials
FREE online support
continuing Glycomics education and research.

Readers have access to hundreds of hours of FREE online
materials in the form of articles, reports, and video clips.
This is a part of the educational effort of
The Endowment for Medical Research, Inc.

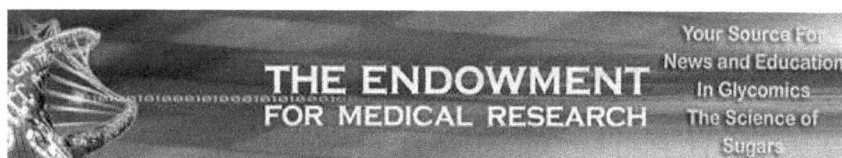

www.endowmentmed.org
Details on booking JC Spencer for lectures
at universities and fund raising events
contact him at jcs@endowmentmed.org

www.GlycoscienceNEWS.com

www.DiabeticHope.com

The author condenses the knowledge of Glycoscience he has gained during the last two decades into three brief sections for the layperson to better understand and for seasoned medical scientists, professors, and researchers to gain new information to better comprehend the depth and importance of this emerging discipline of science.

Chapter One **What are Smart Sugars?**

Chapter Two **Glycomics Holds the Answer to Cancer for Prevention, Diagnosis, and Treatment**

Chapter Three   **Glycoscience Meets Quantum Physics**

Glycoscience is entering mainstream medicine and was declared by MIT as:
"ONE OF THE 10 EMERGING TECHNOLOGIES THAT WILL CHANGE THE WORLD"

Sponsored by

The Endowment for Medical Research Inc, a 501(c)(3) non-profit faith based scientific research, educational, Public Charity
P O Box 73089 - Houston, Texas 77273

www.glycoscienceNEWS.com
or
www.glycoNEWS.com

Profits from the sale of this book are used for continuing education and research of Glycomics

www.ingramcontent.com/pod-product-compliance
Lightning Source LLC
Chambersburg PA
CBHW060514220326
41598CB00025B/3648